Neill Hamilton

FURTHER MATHS REVISION BOOKLET
PURE MATHS 2

COLOURPOINT
EDUCATIONAL

Contents

Revision Exercise 1 3

Revision Exercise 2 8

Revision Exercise 3 13

Revision Exercise 4 19

Revision Exercise 5 25

Answers 29

CCEA GCSE

FURTHER MATHS REVISION BOOKLET
PURE MATHS 2

Neill Hamilton

COLOURPOINT
EDUCATIONAL

Name:

© Neill Hamilton and Colourpoint Creative Ltd 2021

ISBN: 978 1 78073 317 3

First Edition
Second Impression, 2024

Layout and design: April Sky Design
Printed by: GPS Colour Graphics Ltd, Belfast

All rights reserved. No part of this publication may be reproduced, stored in a retrieval system or transmitted in any form or by any means, electronic, mechanical, photocopying, scanning, recording or otherwise, without the prior written permission of the copyright owners and publisher of this book.

The Author

Neill Hamilton will be well known to Mathematics teachers in Northern Ireland. Until his retirement in 2012, he was a teacher of GCSE Mathematics and Additional/Further Mathematics at a Northern Ireland comprehensive school. His previous publications include *Further Mathematics for CCEA GCSE*, and GCSE Mathematics Revision Booklets *M3* and *M4*, also published by Colourpoint.

Dedicated to Arlene, for everything she has done for me, and to Marley who is the best and most loyal friend I could ever have.

Colourpoint Educational
An imprint of Colourpoint Creative Ltd
Colourpoint House
Jubilee Business Park
21 Jubilee Road
Newtownards
County Down
Northern Ireland
BT23 4YH

Tel: 028 9182 0505
E-mail: sales@colourpoint.co.uk
Web site: www.colourpointeducational.com

Publisher's Note: This book has been written to help students preparing for the GCSE Further Mathematics specification from CCEA. While Colourpoint Educational and the authors have taken every care in its production, we are not able to guarantee that the book is completely error-free. Additionally, while the book has been written to closely match the CCEA specification, it is the responsibility of each candidate to satisfy themselves that they have fully met the requirements of the CCEA specification prior to sitting an exam set by that body. For this reason, and because specifications change with time, we strongly advise every candidate to avail of a qualified teacher and to check the contents of the most recent specification for themselves prior to the exam. Colourpoint Creative Ltd therefore cannot be held responsible for any errors or omissions in this book or any consequences thereof.

Revision Exercise 1

1. Write $\dfrac{x^2 - x - 6}{x(x + 2)} + \dfrac{4}{x - 1}$ as a single fraction in its simplest form.

   Answer _____ [4]

2. Expand and simplify the expression $(x + 4)(x - 2)(3x - 5)$

   Answer _____ [3]

3. A function $f(x)$ is defined by $f(x) = x^2 + 6x - 2$
   (a) Use the method of completing the square to rewrite $f(x)$ in the form $(x + a)^2$ where a and b are constants.

   Answer _____ [2]

   (b) Hence find the minimum value of $f(x)$ and the value of x for which it occurs.

   Answer _____ [2]

4. Solve the equations:
   $3x + 2y - z = 15$
   $x - 3y + 2z = 3$
   $2x + y - 3z = 20$

   Answer _____ [8]

5. Solve $3x^2 + 13x - 10 < 0$

   Answer _____ [4]

6. (a) Sketch the graph of $y = \sin x$ for $0° \leq x \leq 360°$

   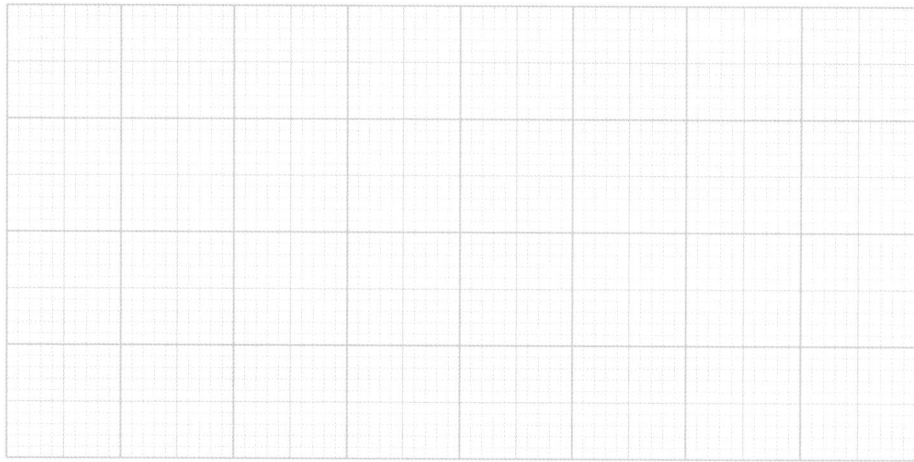

   [1]

(b) Solve the equations
   (i) $\sin x = -0.4$ for $0° \leq x \leq 360°$

   Answer _____ [2]

   (ii) $\sin(2\theta - 15) = -0.4$ for $0° \leq \theta \leq 180°$

   Answer _____ [2]

7. $\log a = x$, $\log b = y$ and $\log c = z$. Find, in terms of x, y and z:

   (a) $\log \dfrac{ab}{c}$

   Answer _____ [1]

   (b) $\log \dfrac{b}{c^2}$

   Answer _____ [1]

   (c) $\log \sqrt{a^3 b}$

   Answer _____ [2]

8. Solve the equation $8^x = 5^{2x-1}$ to 2 decimal places.

Answer _____ [4]

9. Arlene recorded the base lengths, l cm, and the volumes, V cm³, of 5 different solids.
The results are shown in the table below.

Base length l cm	Volume V cm³
4	46.8
5.6	103.16
6.2	131
7.5	205
8	238.5

Arlene believes that a relationship of the form $V = al^n$ exists, where a and n are constants.

(a) Verify that a relationship of the form $V = al^n$ exists by drawing a suitable straight line graph on the grid below. Show clearly the values used, correct to 3 decimal places.
Hence find the values of a and n, correct to 2 decimal places.

Answer _____ [11]

Use the formula $V = al^n$ with your values for a and n to calculate to 2 decimal places:

(b) the volume of a solid with base length 5.1 cm.

Answer _____ [1]

(c) the base length of a solid with volume 260 cm³.
State any assumption that you make.

Answer _____

_____ [2]

Total for revision exercise [50]

Revision Exercise 2

1. Write $\dfrac{y^2 - 8y + 15}{y^2 - 3y} + \dfrac{y - 2}{y + 3}$ as a single fraction in its simplest form.

   Answer _____ [4]

2. Expand and simplify the expression $(4x - 3)^3$

   Answer _____ [3]

3. A function $f(x)$ is defined by $f(x) = x^2 - 3x + 10$
   (a) Use the method of completing the square to rewrite $f(x)$ in the form $(x + a)^2$ where a and b are constants.

   Answer _____ [2]

   (b) Hence find the minimum value of $f(x)$ and the value of x for which it occurs.

   Answer _____ [2]

Revision Exercise 2

4. The equation of a curve is $y = ax^3 + bx^2 + cx - 5$
   Three points on this curve are $(1, -2)$, $(2, 7)$ and $(-3, -98)$.
   (a) Work out the values of a, b and c.

   Answer _____ [9]

   (b) Hence find the values of x where this curve crosses the line $y = 4x - 5$

   Answer _____ [2]

5. Solve $20 - 13x + 2x^2 \geq 0$

   Answer _____ [4]

6. (a) Sketch the graph of $y = \cos x$ for $-360° \leq x \leq 180°$ [4]

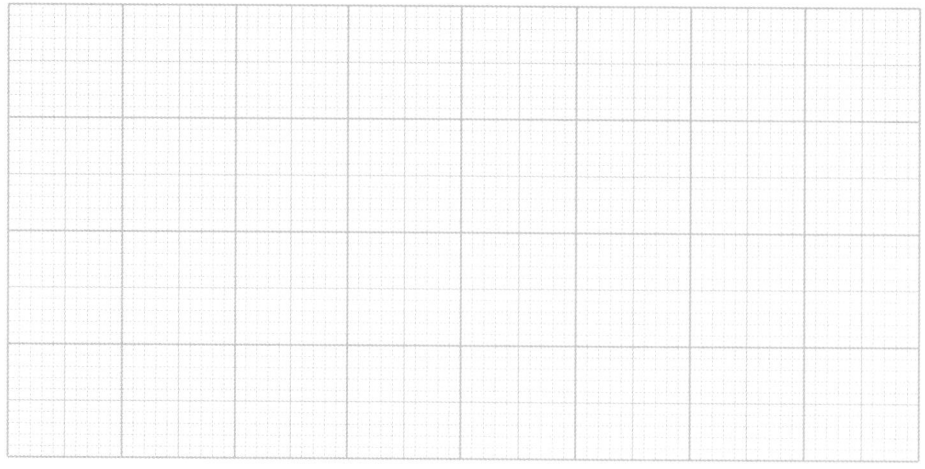

(b) Solve the equation $\cos(\tfrac{2}{5}x - 20) = 0.658$ for $-90° \leq x \leq 180°$

Answer _____ [3]

7. $\log 2 = x$ and $\log 5 = y$
Find, in terms of x and y:
(a) $\log 20$

Answer _____ [2]

(b) $\log 2.5$

Answer _____ [2]

8. Solve the equation $6^{3x+2} = 4^{x-5}$ = to 2 decimal places.

Answer _____ [4]

9. David recorded the number of practice runs, N, and the time taken, T mins, to complete a 5K run for 5 people.
The results are shown in the table below.

Number (N)	Time (T mins)
5	33.6
8	28.25
12	24.3
20	20.1
28	17.7

David believes that a relationship of the form $T = vN^w$ exists, where v and w are constants.

(a) Verify that a relationship of the form $T = vN^w$ exists by drawing a suitable straight line graph on the grid below. Show clearly the values used, correct to 3 decimal places.
Hence find the values of v and w, correct to 2 decimal places.

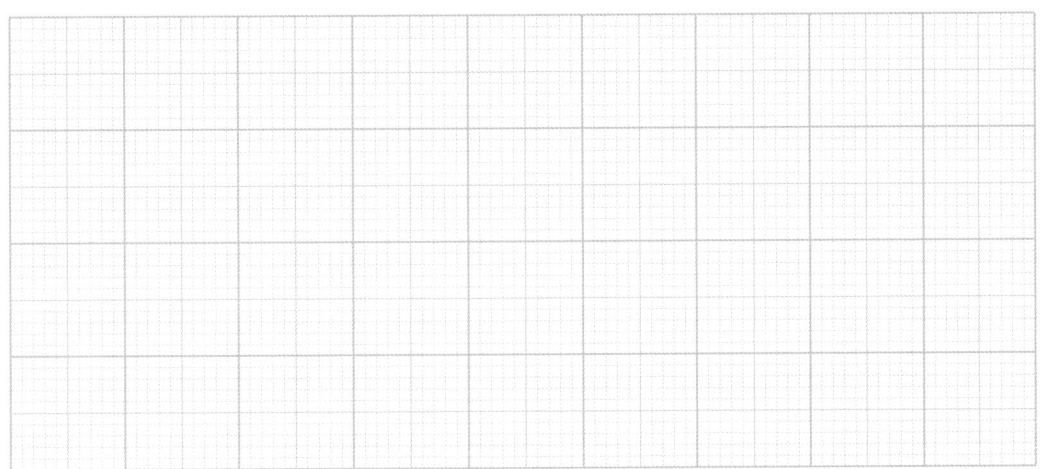

Answer _____ [11]

Use the formula $T = vN^w$ with your values for v and w to calculate:

(b) the time taken, to 1 decimal place, to complete a 5K run after 15 practice runs.

Answer _____ [1]

(c) the number of practice runs needed to complete a 5K run in under 35 mins 24 sec.
State any assumption that you make.

Answer _____

_____ [2]

Total for revision exercise [**50**]

Revision Exercise 3

1. (a) Show that the equation $\dfrac{3t^2 + t - 2}{9t^2 - 4} + \dfrac{2t}{t - 2} = 2$ can be written as $t^2 + 11t + 6 = 0$

   Answer _____ [4]

   (b) Hence solve $\dfrac{3t^2 + t - 2}{9t^2 - 4} + \dfrac{2t}{t - 2} = 2$ giving answers correct to 2 decimal places.

   Answer _____ [1]

2. (a) Expand and simplify the expression $(2x - 1)(x + 2)(x + 3)$

   Answer _____ [2]

   (b) Hence simplify fully the expression $\dfrac{(2x - 1)(x + 2)(x + 3) - 3x(x + 5) + 6}{2x^2 - 2x}$

   Answer _____ [2]

3. A function $f(x)$ is defined by $f(x) = x^2 + x - 11$
   (a) Use the method of completing the square to rewrite $f(x)$ in the form $(x + a)^2$ where a and b are constants.

   Answer _____ [2]

   (b) Hence find the minimum value of $f(x)$ and the value of x for which it occurs.

   Answer _____ [2]

4. On Monday sweets cost x pence each, packets of crisps cost y pence each and chocolate bars cost z pence each in a shop.
   Willow bought 6 sweets, 4 packets of crisps and 2 chocolate bars on Monday for £3.18.
   (a) Show that $3x + 2y + z = 159$

   Answer _____ [1]

   Lily bought 9 sweets, 3 packets of crisps and 12 chocolate bars on Monday. She paid with a £10 note and got £2.08 change.
   (b) Show that $3x + y + 4z = 264$

   Answer _____ [1]

On Tuesday, the packets of crisps were sold at half price and each chocolate bar was sold at 5p less than it was previously. The sweets were the same price.
On Tuesday Cadence bought 4 sweets, 6 packets of crisps and 2 chocolate bars for £2.42.

(c) Show that $4x + 3y + 2z = 252$

Answer _____ [1]

(d) Hence work out the values of x, y and z.

Answer _____ [5]

(e) Quinn had £5 to spend on Monday. He bought the same number of sweets, packets of crisps and chocolate bars and as many of these as possible. Work out his change.

Answer _____ [1]

(f) Rory bought 3 sweets and 2 packets of crisps and a number of chocolate bars on Tuesday. He paid with 3 £1 coins and got 56p change.
How many chocolate bars did he buy?

Answer _____ [1]

5. Solve $3x^2 \leq 15x$

Answer _____ [4]

6. (a) Sketch the graph of $y = \tan x$ for $-90° \leq x \leq 270°$

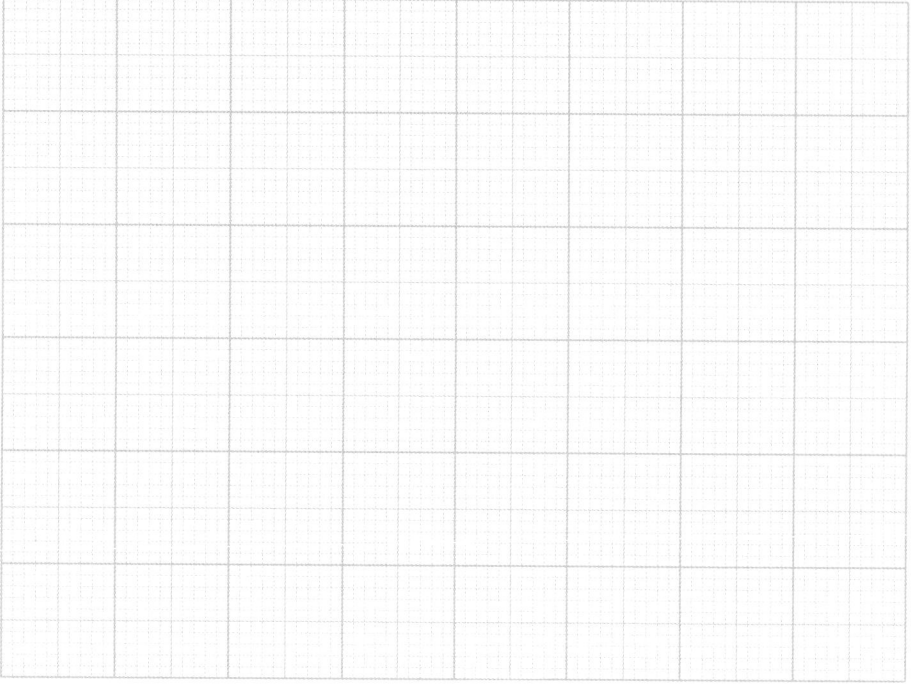

[1]

(b) Solve the equations:
   (i) $\tan x = -3$ for $-90° \leq x \leq 270°$

Answer _____ [1]

   (ii) $\tan(7 - 2\theta) = -3$ for $-90° \leq \theta \leq 45°$

Answer _____ [2]

Revision Exercise 3

7. Find the value of n for which:
   (a) $\log_n 81 = 4$

   Answer _____ [1]

   (b) $\log_2 n = 5$

   Answer _____ [1]

8. Solve the equation $7^{1-\frac{2}{3}x} = 5^{x-2}$ to 2 decimal places.

   Answer _____ [4]

9. The heart–beat rate H beats per minute of a cyclist was recorded M minutes after he finished cycling. The results are shown in the table below.

M minutes	H beats per minute
1.5	167.6
2.5	136.8
4	113.5
7.5	88.3
9	82.2

Colum believes that a relationship of the form $H = vM^q$ exists, where v and q are constants.

(a) Verify that a relationship of the form $H = vM^q$ exists by drawing a suitable straight line graph on the grid below. Show clearly the values used, correct to 3 decimal places.
Hence find the values of v and q, correct to 3 significant figures.

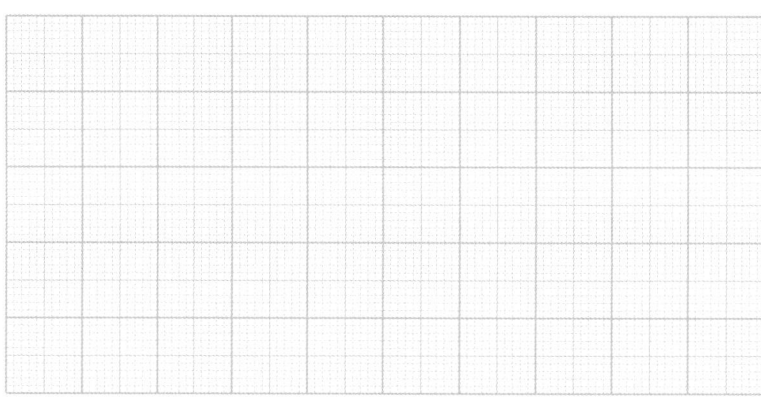

Answer _____ [11]

Use the formula $H = vM^q$ with your values for v and q to calculate to 3 significant figures:

(b) the heartbeat rate after 5 minutes.

Answer _____ [1]

(c) after what time the heartbeat rate is 150 beats per minute.

Answer _____ [1]

Total for revision exercise [50]

Revision Exercise 4

1. **(a)** Show that the equation $\dfrac{v^2 + 3v}{3v + 9} - \dfrac{v}{v - 4} = 1$ can be written as $v^2 - 10v + 12 = 0$

   Answer _____ [4]

   (b) Solve $v^2 - 10v + 12 = 0$ giving your answers correct to 2 decimal places.

   Answer _____ [1]

2. Simplify $(2x - 1)(x + 4)(x - 3) + (3x - 2)(x + 5)$

   Answer _____ [3]

3. Use the method of completing the square to solve the equation $x^2 - 8x + 2 = 0$
   Give your answer in the form $a \pm \sqrt{b}$ where a and b are whole numbers.

   Answer _____ [3]

4. In a theatre:
   - seats in the stalls cost £x each
   - seats in the circle cost £y each
   - seats in the balcony cost £z each

   At one matinee performance 40 seats were sold in the stalls, 25 seats were sold in the circle and 10 seats were sold in the balcony.
   The total price of all these seats was £1335.

   (a) Show that $8x + 5y + 2z = 267$

   Answer _____ [1]

   At one evening performance 56 seats were sold in the stalls, 42 seats were sold in the circle and 84 seats were sold in the balcony.
   The total price of all these seats was £2926.

   (b) Show that $4x + 3y + 6z = 209$

   Answer _____ [1]

   There were two last night performances, at both of which each seat in the stalls cost £4 less than previously, each seat in the circle cost £2 less than previously and the cost of each seat in the balcony was reduced by ⅓ of its previous price.
   At the first last night performance 72 seats were sold in the stalls, 56 seats were sold in the circle and 90 seats were sold in the balcony.
   The total price of all these seats was £2746.

   (c) Show that $36x + 28y + 30z = 1573$

   Answer _____ [1]

(d) Hence work out the values of x, y and z.

Answer _____ [5]

At the second last night performance twice as many seats were sold in the stalls as in the circle and ⅓ more seats were sold in the balcony than in the stalls.
The total price of all these seats was £3312.
(e) How many seats were sold altogether?

Answer _____ [2]

5. A rectangle has length x cm and width $(x - 2)$ cm, as shown in the diagram.

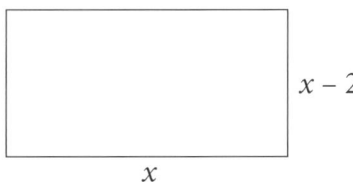

Its area is at least 8 cm².
Find the range of possible values of x.

Answer _____ [4]

6. Solve the equations:
   (a) $\sin x = 0.5$ for $-360° \leq x \leq 0°$

   Answer _____ [1]

   (b) $\sin(4\theta + 10) = 0.5$ for $-90° \leq \theta \leq 0°$

   Answer _____ [2]

7. $\log_3 2 = x$ and $\log_3 5 = y$
   Find, in terms of x and y:
   (a) $\log_3 18$

   Answer _____ [2]

   (b) $\log_3 75$

   Answer _____ [2]

8. Solve the equation $9^{2x-1} = 2^{2-3x}$ to 2 decimal places.

Answer _____ [4]

9. The wind chill index, C, was recorded for different wind speeds, S km/h. The results are shown in the table below.

S km/h	C
8	27.9
11	29.7
19	33.2
46	39.6
75	43.6

Lily believes that a relationship of the form $C = dS^n$ exists, where d and n are constants.

(a) Verify that a relationship of the form $C = dS^n$ exists by drawing a suitable straight line graph on the grid below. Show clearly the values used, correct to 3 decimal places.
Hence find the values of d and n, correct to 1 decimal place.

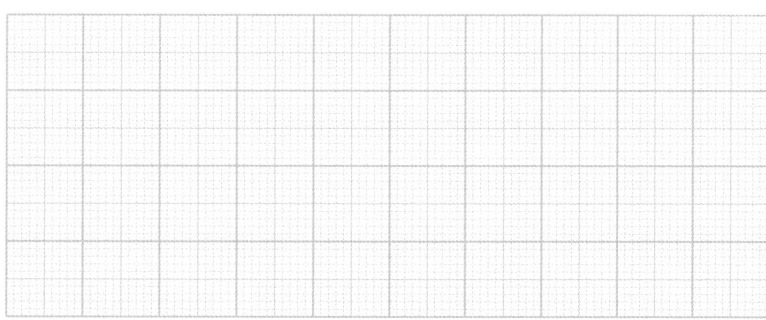

Answer _____ [11]

Use the formula $C = dS^n$ with your values for d and n to calculate, to 1 decimal place:
(b) the wind chill index for a wind speed of 6 km/h and state any assumption that you make.

Answer _____

_____ [2]

(c) the range of wind speeds for a wind chill index that is between 35 and 42.

Answer _____ [1]

Total for revision exercise [50]

Revision Exercise 5

1. Write $\dfrac{3x^2 - 7x - 6}{6x^2 + x - 2} + \dfrac{x - 4}{x}$ as a single fraction in its simplest form.

    Answer _____ [4]

2. (a) Expand and simplify the expression $(2x + 3)^3$

    Answer _____ [3]

    (b) Hence simplify fully $\dfrac{(2x + 3)^3 - 12x(2x + 5) - 2x - 27}{8x^2 - 4x}$

    Answer _____ [3]

3. Use the method of completing the square to solve the equation $x^2 + 5x - 3 = 0$
   Give your answer in the form $a \pm \sqrt{b}$ where a and b are whole numbers.

    Answer _____ [3]

4. Solve the equations:
$2x + 3y - z = 2$
$x - 2y - 4z = -6$
$3x + y - 2z = 8$

Answer _____ [8]

5. There are p packets in a box.
Each packet weighs $(2p + 3)$ grams.
The weight of all the packets is no more than 860 grams.
Find the range of values of p.

Answer _____ [5]

6. Solve the equation $\tan(\tfrac{2}{3}x - 5) + 1 = 0$ for $-90° \leq x \leq 270°$

Answer _____ [4]

7. If $\tfrac{1}{2} \log m = 4 \log n$ write an expression for m in terms of n.

Answer _____ [3]

8. Solve the equation $7^{1-\frac{1}{4}x} = 4^{\frac{2}{3}x}$ to 2 decimal places.

Answer _____ [4]

9. The number of animals, N, in a reserve were recorded after T years.
The results are shown in the table below.

T years	Number of animals N
2	4462
7	1811
11	1308
16	998
24	746

Willow believes that a relationship of the form $N = aT^b$ exists, where a and b are constants.

(a) Verify that a relationship of the form $N = aT^b$ exists by drawing a suitable straight line graph on the grid below. Show clearly the values used, correct to 3 decimal places.
Hence find the values of a to the nearest integer and b correct to 2 decimal places.

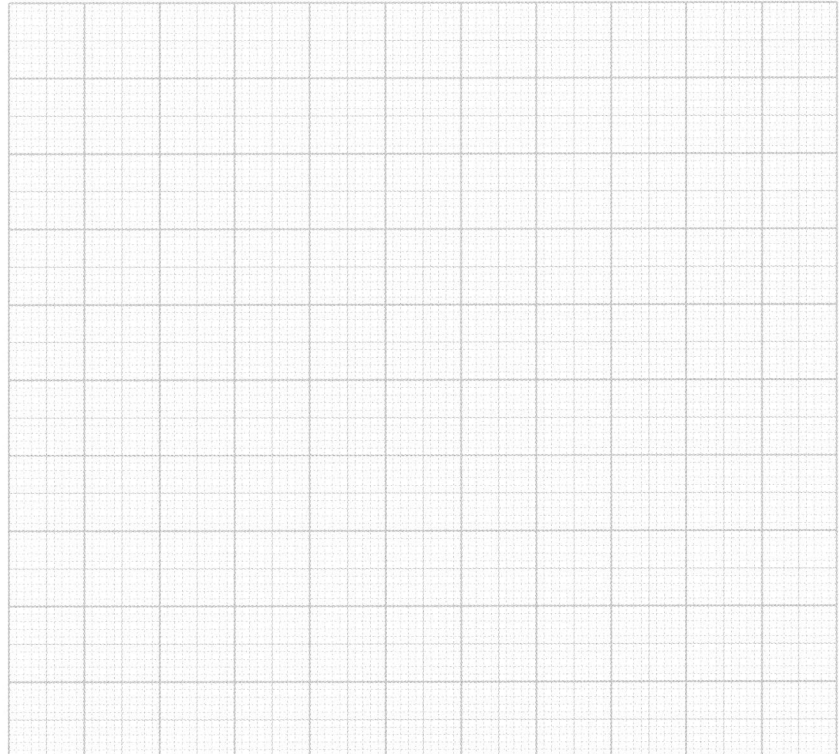

Answer _____ [11]

Use the formula $N = aT^b$ with your values for a and b to calculate:
(b) the number of animals in the reserve after 20 years.

Answer _____ [1]

(c) after how many months the number of animals in the reserve will be less than 2500, correct to the nearest month.

Answer _____ [1]

Total for revision exercise **[50]**

Answers

Revision Exercise 1

1. $\dfrac{(x-3)(x+2)}{x(x+2)} + \dfrac{4}{x-1} = \dfrac{(x-3)}{x} + \dfrac{4}{x-1} = \dfrac{x^2+3}{x(x+1)}$ [M2W2]

2. $(x+4)(x-2) = x^2 + 2x - 8$ [1]
   $(3x-5)(x^2+2x-8) = 3x^3 + x^2 - 34x + 40$ [2]

3. (a) $(x+3)^2 - 9 - 2 = (x+3)^2 - 11$ [2]
   (b) -11 when $x = -3$ [2]

4. $[3x + 2y - z = 15] \times 2$ gives: $6x + 4y - 2z = 30$
   Adding this to $[x - 3y + 2z = 3]$ gives:
   $7x + y = 33$ ...(1)
   $[3x + 2y - z = 15] \times 3$ gives: $9x + 6y - 3z = 45$
   Subtracting $[2x + y - 3z = 20]$ from this gives:
   $7x + 5y = 25$ ...(2)
   Subtracting (2) from (1) gives: $-4y = 8$, so $y = -2$
   Substituting $y = -2$ in (1) gives: $7x - 2 = 33$, so $x = 5$
   Substituting $x = 5$ and $y = -2$ in $x - 3y + 2z = 3$
   gives: $5 + 6 + 2z = 3$, so $z = -4$ [M4W4]

5. $(3x-2)(x+5) < 0$, Answer: $-5 < x < \tfrac{2}{3}$ [M2W2]

6. (a) [1]

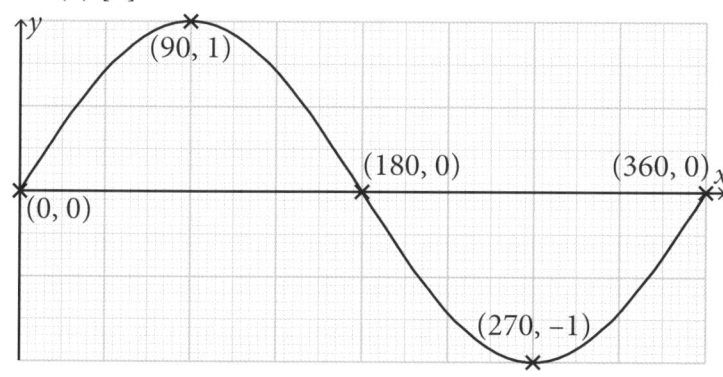

(b) (i) $203.58°$ or $336.42°$ [2] (ii) $2\theta - 15 = 203.58°$
   or $336.42°$; $\theta = 109.29°$ or $175.71°$ [2]

7. (a) $\log a + \log b - \log c = x + y - z$ [1]
   (b) $\log b - \log c^2 = \log b - 2\log c = y - 2z$ [1]
   (c) $\log = \tfrac{3}{2}\log a + \tfrac{1}{2}\log b = \tfrac{3}{2}x + \tfrac{1}{2}y$ [2]

8. $x \log 8 = (2x - 1) \log 5$
   $x \log 8 = 2x \log 5 - \log 5$
   $x \log 8 - 2x \log 5 = -\log 5$
   $x(\log 8 - 2\log 5) = -\log 5$
   $x = \dfrac{-\log 5}{\log 8 - 2\log 5} = 1.41$ [M1W3]

9. (a) $\log V = n \log l + \log a$

$\log l$	$\log V$
0.602	1.670
0.748	2.014
0.792	2.117
0.875	2.312
0.903	2.377

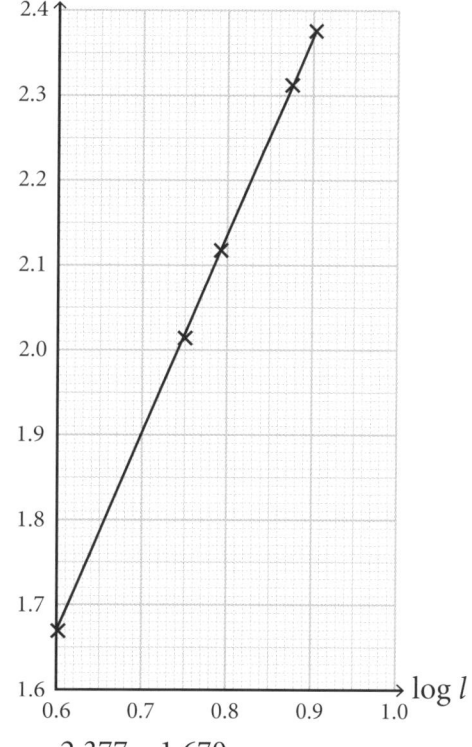

$n = \dfrac{2.377 - 1.670}{0.903 - 0.602} = 2.35$

$V = al^n$, so $46.8 = a(4^{2.35})$, $25.99a = 46.8$, $a = 1.80$
[M4W7]

(b) $V = 1.80\, l^{2.35} = 1.80 \times 5.1^{2.35} = 82.81\ \text{cm}^3$ [1]

(c) $1.80\, l^{2.35} = 260$, so $l^{2.35} = 144.44$, $l = 8.30$ cm
Assume relationship holds for volumes greater than 238.5. [2]

Revision Exercise 2

1. $\dfrac{(y-3)(y-5)}{y(y-3)} - \dfrac{y-2}{y+3} = \dfrac{y-5}{y} - \dfrac{y-2}{y+3} = \dfrac{-15}{y(y+3)}$
   [M2W2]

2. $64x^3 - 144x^2 + 108x - 27$ [3]

3. (a) $(x - \tfrac{3}{2})^2 + 3\tfrac{1}{4}$ [2] (b) $3\tfrac{1}{4}$ when $x = \tfrac{3}{2}$ [2]

4. The equation of a curve is $y = ax^3 + bx^2 + cx - 5$
   Substituting $(1, -2)$ gives $a + b + c - 5 = -2$
   Substituting $(2, 7)$ gives $8a + 4b + 2c - 5 = 7$
   Substituting $(-3, -98)$ gives $-27a + 9b - 3c - 5 = -98$
   (a) $a = 2$, $b = -3$ and $c = 4$
   (b) $2x^3 - 3x^2 + 4x - 5 = 4x - 5$, so $2x^3 - 3x^2 = 0$
   so $x^2(2x - 3)$, so $x = 0$ or $\tfrac{3}{2}$ [M4W5]

5. $(5 - 2x)(4 - x) \geq 0$, Answer: $x \leq \tfrac{5}{2}$ or $x \geq 4$
   [M2W2]

6. (a) [1]

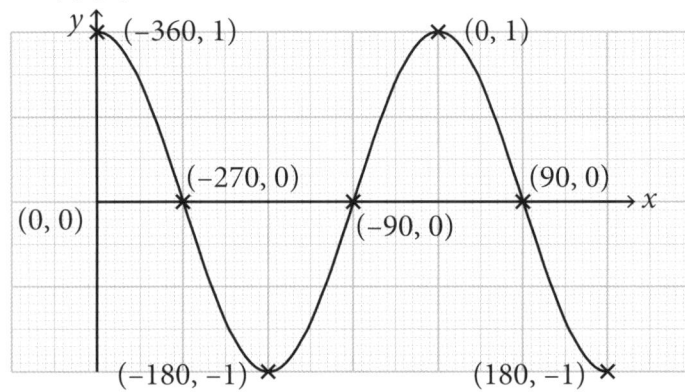

   (b) $\tfrac{2}{5}x - 20 = 48.85$ or -48.85, so $\tfrac{2}{5}x = 68.85$ or -28.85, so $x = 172.13°$ or $-72.13°$ [M1W2]

7. (a) $\log 2^2 \times 5 = 2 \log 2 + \log 5 = 2x + y$ [1]
   (b) $\log \tfrac{5}{2} = \log 5 - \log 2 = y - x$ [1]

8. $(3x + 2) \log 6 = (x - 5) \log 4$
   $3x \log 6 + 2 \log 6 = x \log 4 - 5 \log 4$
   $3x \log 6 - x \log 4 = -5 \log 4 - 2 \log 6$
   $x(3 \log 6 - \log 4) = -5 \log 4 - 2 \log 6$
   $x = \dfrac{-5 \log 4 - \log 4}{3 \log 6 - \log 4} = -2.64$ [M1W3]

9. (a) $\log T = w \log N + \log v$

$\log N$	$\log T$
0.699	1.526
0.903	1.451
1.079	1.386
1.301	1.303
1.447	1.248

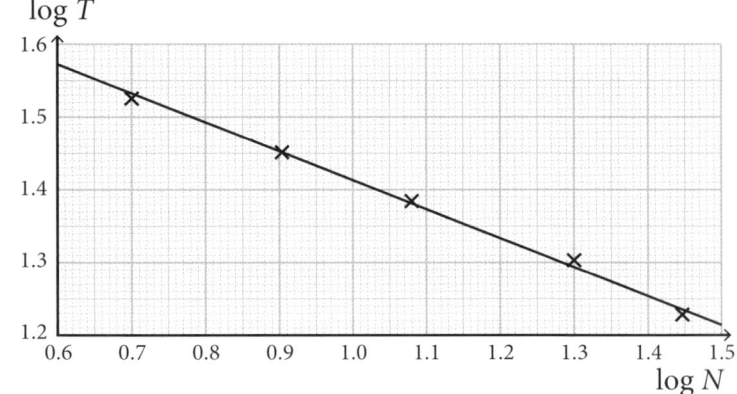

   $w = \dfrac{1.248 - 1.526}{1.447 - 0.699} = -0.37$

   $T = vN^w$, so $33.6 = v(5^{-0.37})$, $v = 60.95$ [M4W7]
   (b) $T = 60.95 N^{-0.37} = 60.95 \times 15^{-0.37} = 22.4$ mins [1]
   (c) $60.95 N^{-0.37} = 35.4$, so $N = 4$. Assume relationship holds for times greater than 33.6 mins. [2]

Answers

Revision Exercise 3

1. $\dfrac{(3t-2)(t+1)}{2t(3t+2)} - \dfrac{2t}{t-2} = \dfrac{t+1}{3t+2} + \dfrac{2t}{t-2}$

   $= \dfrac{(t+1)(t-2) + 2t(3t+2)}{(3t+2)(t-2)}$

   $= \dfrac{t^2 - t - 2 + 6t^2 + 4t}{3t^2 - 4t - 4} = \dfrac{7t^2 + 3t - 2}{3t^2 - 4t - 4} = 2$

   so $7t^2 + 3t - 2 = 6t^2 - 8t - 8$, so $t^2 + 11t + 6 = 0$ [M1W3]

   (b) $t = \dfrac{-11 \pm \sqrt{97}}{2} = -0.58$ or -10.42 [1]

2. (a) $2x^3 + 9x^2 + 7x - 6$ [2]

   (b) $\dfrac{2x^3 + 9x^2 + 7x - 6 - 3x^2 - 15x + 6}{2x^2 - 2x}$

   $= \dfrac{2x^3 + 6x^2 - 8x}{2x^2 - 2x}$

   $= \dfrac{2x(x^2 + 3x - 4)}{2x(x - 1)} = \dfrac{(x+4)(x-1)}{x-1} = x + 4$ [M1W1]

3. (a) $(x + ½)^2 - 11¼$ [2] (b) $-11¼$ when $x = -½$ [2]

4. (a) $[6x + 4y + 2z = 318] \div 2$
   gives $3x + 2y + z = 159$ [1]
   (b) $[9x + 3y + 12z = 792] \div 3$
   gives $3x + y + 4z = 264$ [1]
   (c) $4x + 6(½y) + 2(z - 5) = 242$,
   so $4x + 3y + 2z - 10 = 242$,
   so $4x + 3y + 2z = 252$ [1]
   (d) $x = 18$, $y = 30$, $z = 45$ [M1W4]
   (e) $18 + 30 + 45 = 93$, $500 \div 93 = 5$, $5 \times 93 = 465$
   $500 - 465 = 35$p change [1]
   (f) $54 + 30 + 40n = 244$ so $n = 4$ [1]

5. $-15x \le 0$ so $3x(x - 5) \le 0$, Answer: $0 \le x \le 5$ [M2W2]

6. (a) [1]

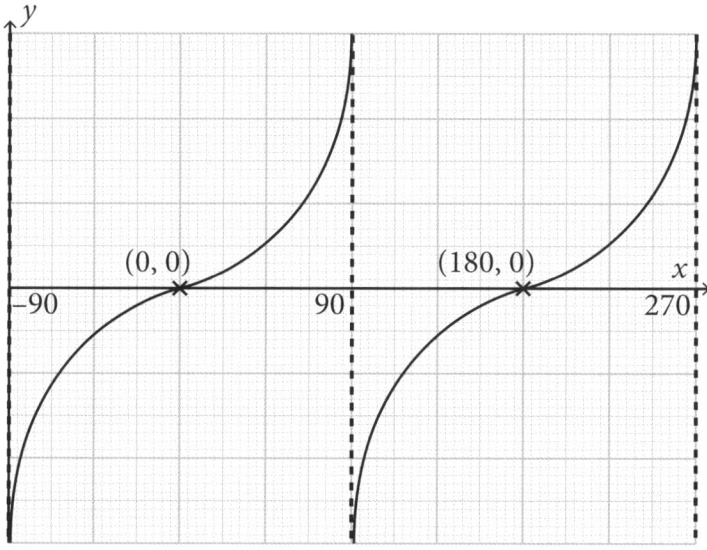

   (b) (i) $108.43°$ or $-71.57°$ [1] (ii) $7 - 2\theta = 108.43°$ or $-71.57°$, so $\theta = -50.72°$ or $39.29°$ [2]

7. (a) $n^4 = 81$, so $n = 3$ [1] (b) $n = 2^5 = 32$ [1]

8. $(1 - \tfrac{2}{3}x) \log 7 = (x - 2) \log 5$
   $\log 7 - 0.563x = 0.699x - 2 \log 5$
   $1.262x = \log 7 + 2 \log 5$, so $x = 1.78$ [M1W3]

9. (a) $\log H = q \log M + \log v$

log T	log H
0.176	2.224
0.398	2.136
0.602	2.055
0.875	1.946
0.954	1.915

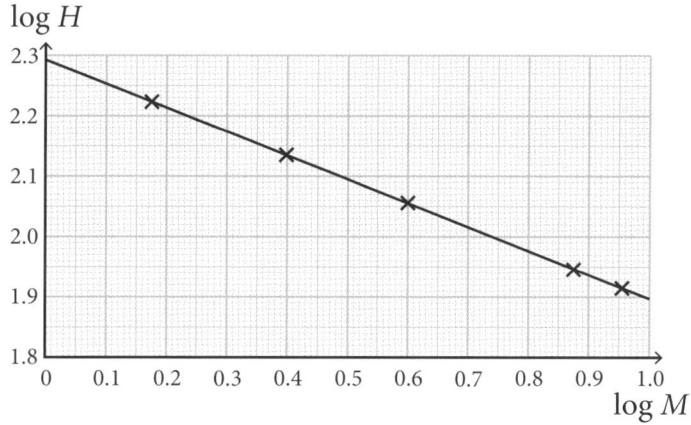

   $q = -0.397$, $v = 197$ [M4W7]
   (b) 104 [1] (c) 1.98 [1]

Revision Exercise 4

1. (a) $\dfrac{v(v+3)}{3(v+3)} - \dfrac{v}{v-4} = \dfrac{v}{3} - \dfrac{v}{v-4} = \dfrac{v^2 - 4v - 3v}{3(v-4)}$

   So $\dfrac{v^2 - 7v}{3v - 12} = 1$, so $v^2 - 7v = 3v - 12$,

   so $v^2 - 10v + 12 = 0$ [M1W3]

   (b) $v = \dfrac{10 \pm \sqrt{100 - 48}}{2}$, Answer: 8.61 or 1.39 [1]

2. $2x^3 + x^2 - 25x + 12 + 3x^2 + 13x - 10$
   $= 2x^3 + 4x^2 - 12x + 2$ [3]

3. $3(x-4)^2 - 16 + 2 = 0$ so $(x-4)^2 = 14$,
   so $x - 4 = \pm\sqrt{14}$
   Answer: $x = 4 \pm \sqrt{14}$ [3]

4. (a) $[40x + 25y + 10z = 1335] \div 5$
   $= 8x + 5y + 2z = 267$ [1]
   (b) $[56x + 42y + 84z = 2926] \div 14$
   $= 4x + 3y + 6z = 209$ [1]
   (c) $72(x - 4) + 56(y - 2) + 90(\tfrac{2}{3}z) = 2746$
   $[72x + 56y + 60z = 3146] \div 2$
   $= 36x + 28y + 30z = 1573$ [1]
   (d) $x = 20, y = 16, z = 13.50$ [M1W4]
   (e) Let n be the number of seats sold in the circle. Then $2n$ are the number of seats sold in the stalls.
   $\tfrac{1}{3}$ of $2n = \tfrac{2}{3}n$ and $2n + \tfrac{2}{3}n = \tfrac{8n}{3}$
   so $\tfrac{8n}{3}$ are the number of seats sold in the balcony,
   so $2n(20) + n(16) + \tfrac{8n}{3}(13.50) = 3312$
   so $40n + 16n + 36n = 3312$
   so $92n = 3312$, so $n = 36$
   Total $= 72 + 36 + 96 = 204$ [2]

5. Area $= x(x - 2) = x^2 - 2x$
   $x^2 - 2x \geq 8$, so $x^2 - 2x - 8 \geq 0$, $(x-4)(x+2) \geq 0$
   So $x \geq 4$ or $x \leq -2$ but $x \leq -2$ is impossible as length must be positive. So answer: $x \geq 4$ [M2W2]

6. (a) $-210°$ or $-330°$ [1]
   (b) $4\theta + 10 = -210°$ or $-330°$,
   so $\theta = -55°$ or $-85°$ [2]

7. (a) $\log_3 2 \times 3^2 = \log_3 2 + 2\log_3 3 = x + 2$ [M1W1]
   (b) $\log_3 3 \times 5^2 = \log_3 3 + 2\log_3 5 = 1 + 2y$ [M1W1]

8. $(2x - 1)\log 9 = (2 - 3x)\log 2$
   so $1.908x - 0.954 = 0.602 - 0.903x$
   so $2.811x = 1.556$
   giving $x = 0.55$ [M1W3]

9. (a) $\log C = n \log S + \log d$

log S	log C
0.903	1.446
1.041	1.473
1.279	1.521
1.663	1.598
1.875	1.639

   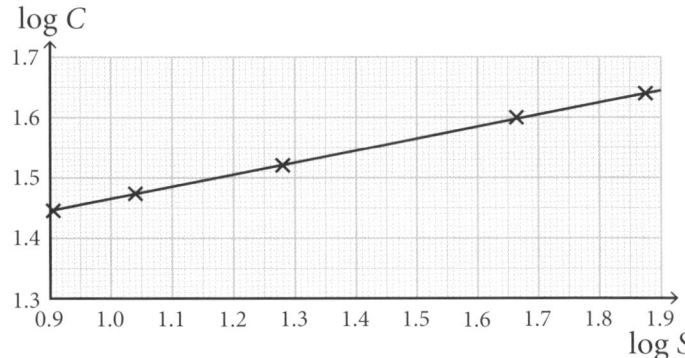

   $n = 0.2, d = 18.4$ [M4W7]
   (b) 26.3 Assume relationship holds for wind speeds less than 8. [2]
   (c) 24.9 to 62.0 km/h [1]

Answers

Revision Exercise 5

1. $\dfrac{(3x+2)(x-3)}{(3x+2)(2x-1)} = \dfrac{(x-3)}{(2x-1)} - \dfrac{(x-4)}{x}$

   $= \dfrac{x(x-3) - (x-4)(2x-1)}{x(2x-1)} = \dfrac{-x^2 + 6x - 4}{x(2x-1)}$ [M1W3]

2. (a) $8x^3 + 36x^2 + 54x + 27$ [3]

   (b) $\dfrac{8x^3 + 36x^2 + 54x + 27 - 24x^2 - 60x - 2x - 27}{8x^2 - 4x}$

   $= \dfrac{8x^3 + 12x^2 - 8x}{8x^2 - 4x} = \dfrac{4x(2x^2 + 3x - 2)}{4x(2x-1)}$

   $= \dfrac{4x(2x-1)(x+2)}{4x(2x-1)} = x + 2$ [3]

3. $(x + 5/2)^2 - 25/4 - 3 = 0$, so $(x + 5/2)^2 = 37/4$, so

   $x + 5/2 = \dfrac{\pm\sqrt{37}}{2}$, so $x = \dfrac{-5 \pm \sqrt{37}}{2}$ [3]

4. $x = 6$, $y = -2$, $z = 4$ [M4W4]

5. Total weight = $p(2p + 3) = 2p^2 + 3p$
   $2p^2 + 3p \leq 860$, so $2p^2 + 3p - 860 \leq 0$, so
   $(2p + 43)(p - 20) \leq 0$, so $-43/2 \leq p \leq 20$
   But p must be positive and a whole number, so
   Answer: $0 < p \leq 20$ [M3W2]

6. $\tan(2/3\,x - 5) = -1$, so $2/3\,x - 5 = 135$ or -45
   Answer: $x = 210°$ or $-60°$ [M2W2]

7. $\log m^{1/2} = \log n^4$, so $m^{1/2} = n^4$, so $m = n^8$ [M1W2]

8. $(1 - 1/4\,x) \log 7 = 2/3\,x \log 4$
   so $0.845 - 0.211x = 0.401x$
   so $0.845 = 0.612x$
   giving $x = 1.38$ [M1W3]

9. (a) $\log N = b \log T + \log a$

$\log T$	$\log N$
0.301	3.650
0.845	3.258
1.041	3.117
1.204	2.999
1.380	2.873

   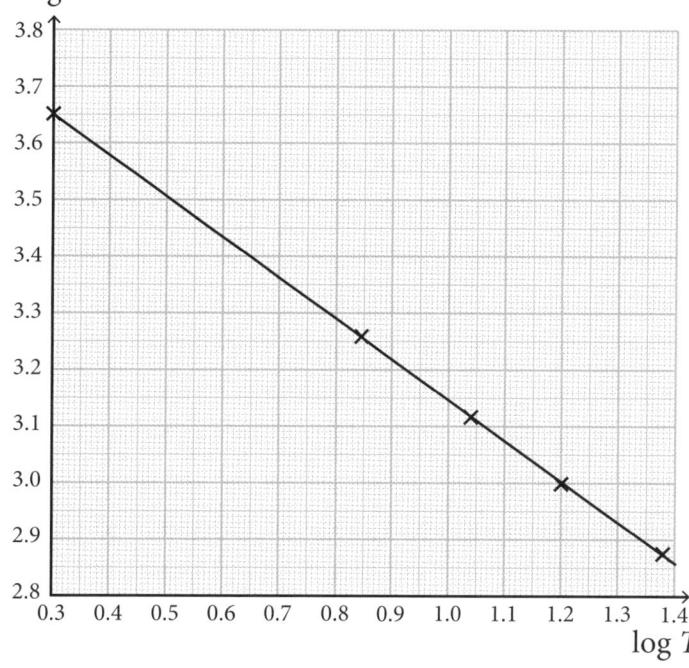

   $b = -0.72$, $a = 7351$ [M4W7]
   (b) 850 or 851 [1]
   (c) 54 months (to the nearest month) [1]

Meeting the requirements of the CCEA GCSE Further Mathematics specification, this is one of four revision booklets that address the course. This workbook is set out in the form of five revision tests and covers all of the elements of Pure Mathematics with the exception of matrices and calculus, which are covered by the workbook *Pure Maths 1*.

These valuable questions were specially commissioned for the booklet and are not from past papers. Full answers are included at the rear and contain not only the final answer but, where appropriate, an indication of the process required to reach the given solution. The workbook has been through a meticulous quality assurance process by a GCSE Mathematics expert.

What workbooks do I need?

Students sitting CCEA GCSE Further Mathematics will usually need four workbooks – *Pure Maths 1* and *Pure Maths 2*, plus the *Mechanics* workbook and the *Statistics* workbook.

£3.99